Machines and Motion

Pulleys

by Erika S. Manley

Bullfrog Books

Ideas for Parents and Teachers

Bullfrog Books let children practice reading informational text at the earliest reading levels. Repetition, familiar words, and photo labels support early readers.

Before Reading

- Discuss the cover photo. What does it tell them?
- Look at the picture glossary together. Read and discuss the words.

Read the Book

- "Walk" through the book and look at the photos. Let the child ask questions. Point out the photo labels.
- Read the book to the child, or have him or her read independently.

After Reading

- Prompt the child to think more. Ask: Pulleys are everywhere. Where do you see them? What do they help you accomplish?

Bullfrog Books are published by Jump!
5357 Penn Avenue South
Minneapolis, MN 55419
www.jumplibrary.com

Library of Congress Cataloging-in-Publication Data

Names: Manley, Erika S., author.
Title: Pulleys / by Erika S. Manley.
Description: Minneapolis, MN : Jump!, Inc., [2018]
Series: Machines and motion
"Bullfrog Books are published by Jump!"
Audience: Ages 5–8. | Audience: K to grade 3.
Includes bibliographical references and index.
Identifiers: LCCN 2017054170 (print)
LCCN 2017054646 (ebook)
ISBN 9781624968563 (e-book)
ISBN 9781624968549 (hardcover : alk. paper)
ISBN 9781624968556 (pbk.)
Subjects: LCSH: Pulleys—Juvenile literature.
Simple machines—Juvenile literature.
Classification: LCC TJ1338 (ebook)
LCC TJ1338 .M31975 2018 (print) | DDC 621.8/11—dc23
LC record available at https://lccn.loc.gov/2017054170

Editor: Kristine Spanier
Book Designer: Molly Ballanger

Photo Credits: Fotofermer/Shutterstock, cover, 12; Kittichai/Shutterstock, 1; MartinMoizis/Shutterstock, 3; andresr/Shutterstock, 4; Have a nice day Photo/Shutterstock, 5, 23tr; gaspr13/Shutterstock, 6–7, 23br; yogesh _ more/Shutterstock, 8–9, 23tl; Tatiana Bobkova/Shutterstock, 10–11, 23bl; PacoRomero/iStock, 12; Blend Images/Superstock, 13; Blend Images/Superstock, 14–15; Tetra Images/Alamy, 16–17; Anetlanda/iStock, 17; Dimedrol68/Shutterstock, 18; Dmitri Ma/Shutterstock, 19; nullplus/iStock, 20–21, igorstevanovic/Shutterstock, 24.

Printed in the United States of America at Corporate Graphics in North Mankato, Minnesota.

Table of Contents

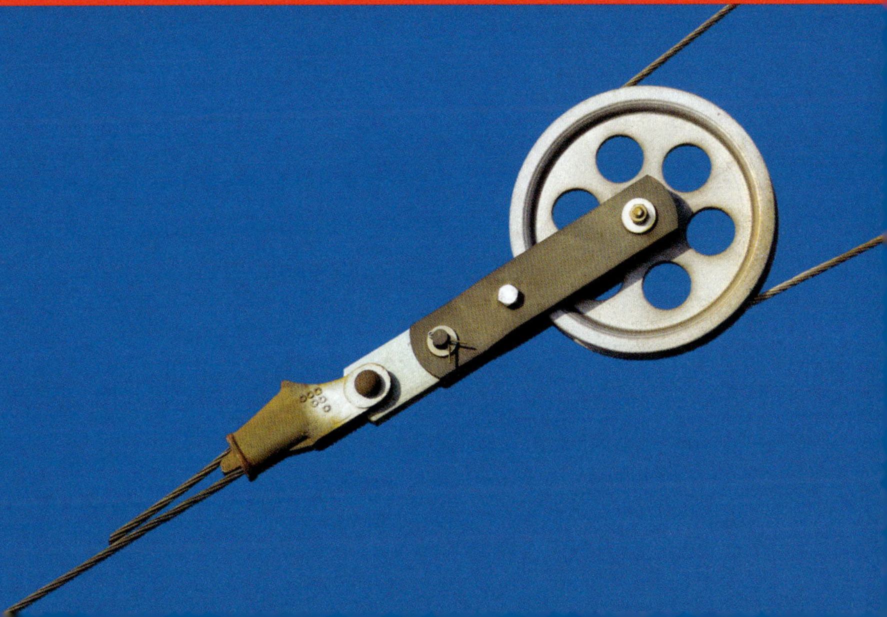

Pull and Lift

What is a pulley?

It is a simple machine.
It helps us lift things.

It has a wheel.

The wheel has a groove in it.

A rope fits in the groove.

groove

axle

What holds the wheel?

An axle.

The line moves.

It lifts the load.

pulley
load

Ann climbs.

She uses a pulley.

She lifts her weight.

A zip line uses one.

Whee!

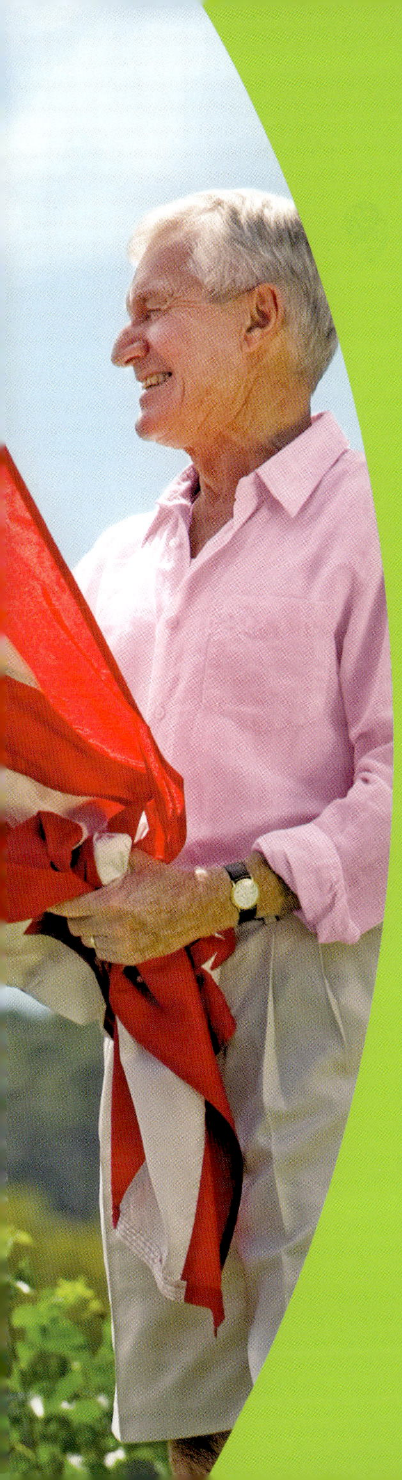

Pat raises a flag.

How?

She pulls on the rope.

Dan's room is bright.

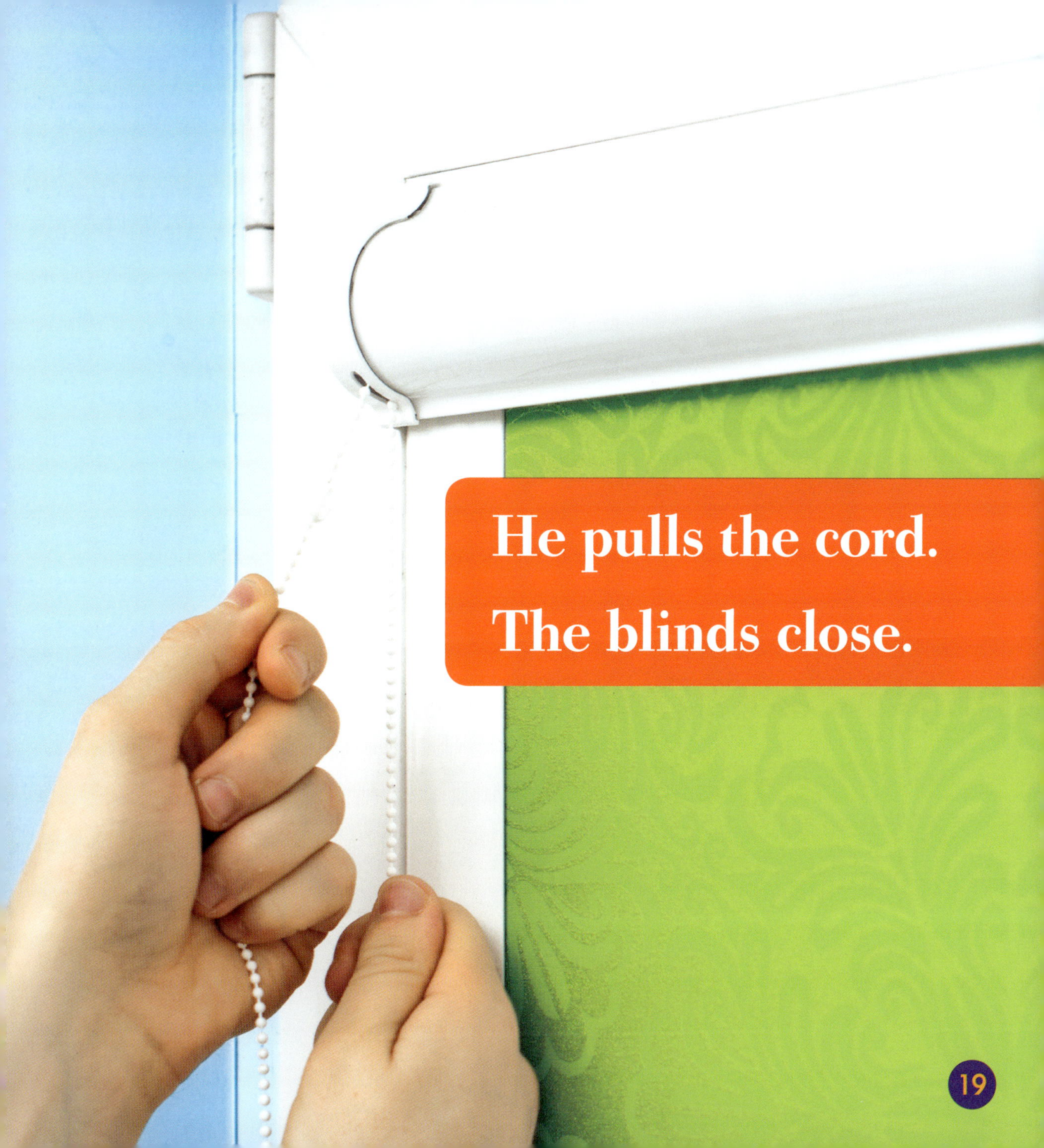

He pulls the cord.
The blinds close.

Need to lift something?

Pulleys can help!

pulley

Build a Pulley

Pulleys help us lift things. Build your own! Have an adult help you with this simple activity.

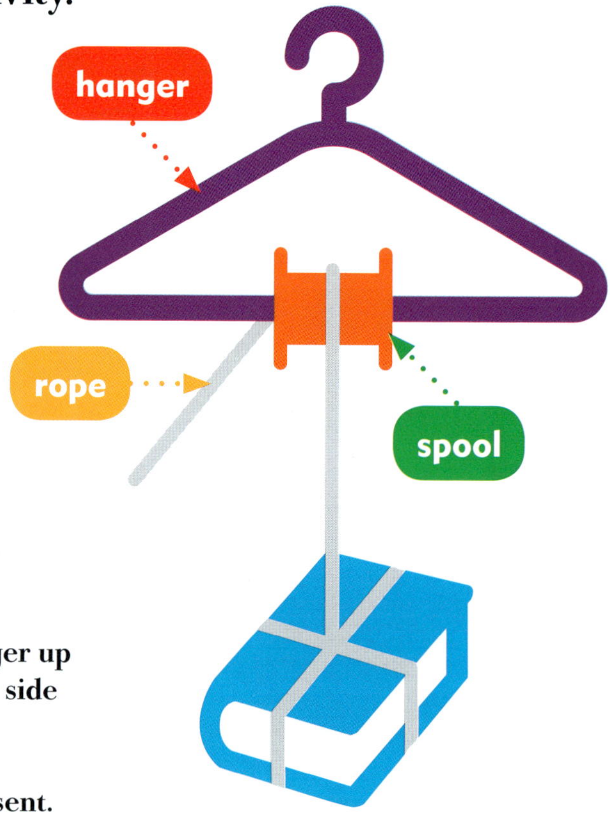

You will need:

- thread spool
- wire hanger
- wire cutter
- pliers
- fine rope (or twine)
- book

Directions:

1. Use a wire cutter to cut the hanger in the middle at the bottom.
2. Slip the spool through the newly cut ends of the hanger. The cut edges should go through either side of the spool.
3. Use the pliers to bend the wire hanger up one inch (2.5 centimeters) on either side of the spool to hold it in place.
4. Wrap the rope around the book on all four sides like a ribbon on a present.
5. Tie a knot at the top to hold the book securely in place.
6. Hang the hanger and pulley system on a closet rod.
7. Lay the book on the closet floor.
8. Glide the opposite end of the rope (non-book end) around the spool.
9. Pull on the rope and watch the book rise!

Picture Glossary

axle
A bar on which a wheel turns.

pulley
A wheel with a grooved rim around which a rope or chain can run.

load
The weight being lifted or moved.

simple machine
A tool used to make work easier, such as an inclined plane, lever, pulley, screw, wedge, or wheel and axle.

Index

To Learn More

Learning more is as easy as 1, 2, 3.

1) Go to www.factsurfer.com
2) Enter "pulleys" into the search box.
3) Click the "Surf" button to see a list of websites.

With factsurfer.com, finding more information is just a click away.